Stormy Weather

written by Dona Smith
illustrated by John Hovell

McGraw-Hill School Division
New York Farmington

The weather is always changing. Sometimes the change is gradual. White clouds drift across a blue sky, or the wind slowly picks up speed. But sometimes the weather can change a lot. And when stormy weather suddenly strikes, it can be big news.

A big storm is coming. It's a cyclone, the biggest kind of storm. Cyclones are also called hurricanes. Other people call these super storms typhoons. In Australia they're called willy-willies. That name may sound silly, but be prepared. There's nothing funny about what cyclones can do.

A cyclone forms over warm ocean waters. It usually begins with heavy rains. Then the winds begin to blow. They can reach up to 200 miles an hour. These winds can smash houses and snap trees in half. The high winds also make the ocean's water roar ashore, flooding roads and buildings. This stormy weather can go on for hours.

Sometimes, the cyclone suddenly stops. A little sunlight might even peek through. This calm spot is called the eye of the storm. Don't be fooled. Once the eye passes, the winds and rain come back again.

England News

MAY 1950

Tornado Plucks Chicken

Today a tornado plucked some chickens in England. The wind pulled their feathers off. They were not happy, but they were alive.

A tornado is a funnel-shaped, gray cloud that reaches up into the sky. Sometimes it is called a twister. That is because it is made of warm air that spins around. The wind speed of a tornado can reach 300 miles an hour.

Tornadoes have also caused some other strange things to happen.

In 1940 in Russia, a tornado made money fall from the sky. The tornado picked up an old treasure chest. The chest broke open in the air and hundreds of gold coins fell out.

Looks like the Empire State Building in New York City has been struck by lightning again. One year it was struck forty times! This proves that lightning can strike twice in the same place.

Why do we see lightning before we hear thunder?

Thunder and lightning happen at the same time, but far above you. You see the lightning snake across the sky. After a brief time, you hear thunder. That is because light travels faster than sound. When a storm is very close, the lightning and thunder are close together, too. When a storm moves away, the time between the lightning and thunder grows.

A hail stone is a chunk of ice. The ice melts and freezes and melts and freezes many times as it falls to Earth. When it lands it is as hard as a rock.

What was the strangest hail?

Once a giant hailstone fell in Mississippi. Inside was a frozen turtle. Winds had lifted the turtle high into the sky. Then ice covered it. That kind of hail is not seen frequently!

On this day almost two feet of snow fell. It snowed everywhere along the East Coast, from Maine to Washington, D.C. People in cities were stranded in trains, carriages, and buildings. Houses were buried!

There have been many other periods when it rained fish and frogs and other strange things. Red rain fell in Switzerland in 1857. It was red because the rain mixed with dust from a desert 1,850 miles (3,000 km) away. The dust had been carried by the wind.

In May 1881, it rained crabs over England. The winds of a twister had sucked the animals into the air. Later, the crabs showered down on land. Everyone wanted to avoid them.

There was a report that herring rained down in Scotland in 1921. In other places, snakes have fallen out of the sky. It has rained snails, too.

People talk about it raining cats and dogs. No one claims to have actually seen it themselves. Maybe some day somebody will.

Are there more strange stories about weather? Yes, there are many more stories. Have you ever seen it rain when it is sunny? The rain can be coming down from a cloud above you. The sun is shining beside the cloud.

Think about this next time it rains and you have to stay inside. Is it just water falling from the sky— or is it crabs or snakes or frogs!?